HERESY AT MATH-SCI. JUNCTION
Basic Mathematics in Physics Theory

Dean LeRoy Sinclair, Ph.D.

Author's Preface

This book is a result of the writer's rather cynical view of the exalted ideas that seem to exist in the sciences as to the value of mathematical models. Rightly or wrongly, he considers that mathematics should be considered as a servant, used to simplify thinking about science and perhaps to organize facts for better understanding. In his opinion, scientists have often come to consider that, somehow, the mathematical models were the controllers of reality rather than approximators.

Ironically, he uses a simple mathematical model as a suspected "controller of existence in his own version of a " Theory of Everything" which is currently called the "Oscillators-in-a-Substance Model of Existence.

It is hoped that this little book will furnish some ideas upon which more accomplished scientists and mathematicians can build.

Much of this material is taken from his other publications, rearranged here for more emphasis on the mathematical aspects and possible errors that may result from overdependence on mathematical modelling.

Essentially, the caveat that is intended here can be summed up as, "A mathematical model is only as valid as the assumptions upon which it is based."

Introduction,

A quote attributed to Albert Einstein is "Mathematics is the reality." Another statement that is often heard is "Mathematics is the Queen of Sciences." Looking at what has been happening in science, particularly in Physics Theory, one may wonder if The "Queen of Sciences" may have changed from a role as a servant to a dictatoress who may often lead astray as she has her rules, her statements and her confusions, yet she is always considered to be right.

Pre-Einstein, scientists tended to collect data, then develop mathematical models to correlate the data. Using these models, they would check for further implications then see if these implicates led to useful results.   Post-Einstein there seems to be a tendency to start with a preconceived model, and, often very expensively, seek some data which will support the model.

One of the less-expensive of these "Expeditions-Following -the-Queen" is the concept of "Black Holes in Space".

A man by the name of Schwarzschild found in some of Einstein's work an idea that there could be "Black Holes" in space.

Astronomers looking for them, found that data having to do with the regions about the centers of mass of galaxies could be interpreted as supporting a "Black Hole" in the center of each Galaxy. They never seem to have considered the more proletarian explanation, that the data actually simply supports the well known fact that the center of mass of any object, will operate as if all of the mass of the object were concentrated at a point. Papers, books, lectures and careers have been based on the more intriguing-- but simply fanciful-- concept of the "Black Hole."

Much worse in terms of wasted time and money is the "Standard Model of Particle Physics". This highly touted and widely believed "House of Cards" started with the idea smashing atoms together could release the fundamental particles that made them up. It seems to have not even been suspected that smashing things together might simply create new and different things from whatever was started with.

  After a few years, the number of "fundamental particles" became so huge that it was decided to try to organize them with some sort of lesser number of units. At about this time, a couple of scientists published what I'll always suspect was clever satire on the tendency of scientists to make things as complicated as possible, in which they proposed a unit which they called a Quark proposing that three forms of this hypothetical unit could be used to explain the situation, and, of course, complicate it further. If the concepts were meant as satire, the scientific community missed the point completely and adopted the suggestion totally, expanding the three Quarks to Six, and the race was on to find more data which could fit into this picture. What resulted, is, in my opinion, one of the most ridiculous fiascos of science: some 40 years of spending billions of dollars chasing a phantom to complete a pattern. This was the search for the "Higgs Boson."

Finally, in 2013, the scientists found some signals which their theorists said could be in the right place for the Higgs" and it was trumpeted that the search had, at last, ended successfully.

In the meantime, this writer, starting in the Spring of 2004 from a consideration of the implications of the Speed of Light in Information Theory had, quite accidentally, developed a much simpler model of existence, called, at the present time, the "Oscillators in a Substance Model."

If this newer model--which actually could have been put together a Century ago--if valid, then the Standard Model is a tangled "House of Cards." We cannot have it both ways.
Of course, it may be that neither of the above theories is truly valid and there be better explanation than either.

One may note the Higgs Concept of a "Field" and a basic particle has correspondences to the "OS" approach. OS postulates a basic "Substance" and a continuously functioning, very high frequency "Control Oscillator" which is continuously organizing and supporting our "Bubble Universe" and, more than likely a number of others, one guess would be some 13 more associated with one 'Control Oscillator." The parallels are obvious, but the details are totally different. Standard Model has fields layered upon fields as basic,The OS model would say that the "Fields" would represent descriptions of disturbance patterns in an overall substance. This is a reversal as to which is fundamental. OS produces a definition for "Mass" which shows that "Mass"-- at least as an instantaneous attribute-- is a necessity for existence, Standard Model has "Massless Gauge bosons...."

In any case, there is reason to reconsider the direction science has taken in doing research.

To me, OS is a "data-driven" model. Standard Model is much the reverse, Date interpretation is driven by the model.

In the book, <u>Oscillators in a Substance Model: A Physicists' Grail and an Alexander's Sword</u>, the writer included several short articles which dealt with basic mathematics. On rereading the book and noting that in that context the material is somewhat supplementary and perhaps superfluous, it has been decided to publish that material separately with additional comment and with a somewhat different audience in mind. This little treatise is the result .

The writer's mathematics training is basic, an undergraduate mathematics major thru differential equations, set theory and non-Euclidian geometry. In many cases he may be "reinventing the wheel," it is hoped that better mathematicians can correct any misapprehensions and/or fallacies of reasoning which appear.

Reality and Mathematical Definitions

We consider mathematics as a model for reality. Perhaps it would be worthwhile to consider how certain mathematical definitions fit as we look at the physical world.

First, let us look at the concept of Zero, usually considered as the symbol of nothingness. However, is this true? Zero is the starting point for counting, the starting point for any journey, the crossing for the Cartesian axes of most conventional graphs. It might be better to say that the symbol, "Zero," is actually the symbol for the fact of existence. In the real world, the starting "point" may be of any size, any shape. Zero, then. is not without existence, without dimension, rather it is the symbol of the very first dimension, the Dimension of Existence.

What then of the number, "One? " That's simple, its the "counting number." However, isn't it much more than that? It is the symbol of wholeness. it represents a whole starting point, a whole line, which actually has to be made up of two starting points, a whole surface, made up of at least three starting points,... Hence, the number one may represent many things.

If we attach a sign to the number, implying a motion, then +l, represents the motion of a whole starting unit one unit to the right, or possibly up, or forward. We say that one times one times one equals one' but.we always assume that there is a positive value attached to the one, so if the first one represents a motion of one space to the right, the next one represents the motion of the first "one" upward, and the third one represents the motion of the second generated one a unit forward. Therefore, if we attach the positive notation to "one" which, by convention we do, one times one times one actually means one cube generated to the right, above and forward from the origin of a set of axes by a set of motions which are actually counter clockwise,

Looking at "One" as the symbol of wholeness has many uses. One interesting one arises is one looks, for instance, at figuring a maximum frequency for our particular Universe.   The equation for the movement of Energy by electromagnetic radiation is E=hu, where "h" is Planck's constant and "u" is cycles per some unit of time.   If we place E equal to one Energy unit in any set of units, the maximum frequency, expressed in that set of units, will be seen to be "**1/h**" . This could be the "high frequency cut off" for communication.

(One correspondent criticized the consideration of "1/h" as a frequency, stating that "1/h" would be the "reciprocal of 'h'" Mathematically, this is what it looks like. However, my correspondent did not realize that, in this case, the symbol for "One" did not represent a "reciprocal" but one "whole unit of energy" with the appropriate unit designations to match whatever "h" was being measured in. This is an example of how simple math definition and usage can mislead, when applied to an attempt at modeling "reality.")

At the other end of the scale is the symbol of "Infinity: the Number Beyond All Numbers." For mathematicians this is a perfectly good definition; but in the real world we perhaps should use a ,more practical definition.
Does it really make sense to say that we can measure mass of something moving with relation to us up to a velocity as near the Speed of Light as we care to but consider that the mass will become "Infinite," meaning "without limit" at the speed of light?
 Is the darkness just beyond the flashlight beam a void? Couldn't that darkness be considered an Infinity? We can't see into it, we can only make a guess as to what is beyond the beam.
  In the practical world we probably should consider the concept of "Infinity" as representing simply the point just beyond the last point that we can measure with the instruments at hand, the number beyond where we stopped counting, for whatever reason.
Be it not practical to consider the infinity symbol as being the

symbol for "Unknown Under Current Conditions?"
This writer feels that, were the above ideas taken into consideration, the theories of physical science would fit more closely to reality than do current theories which are usually considered as being best expressed in differential equations. Differential; equations are expressed in simple numbers, considered, if no sign be designated, to have positive signs.

The following section is made up of two articles from the book mentioned earlier-- two related articles which which appear to be so simple as to be considering obvious ideas, yet, some of the points made do not seem to be considered by the typical scientist. In fact, the discussion of "signed number sets as roots of signed numbers" may be unique to this writer.

On Signed Numbers The +, - Operators

The signs, + , and, - , are used throughout mathematics and science in a number of ways with several meanings which are often not carefully considered. In somewhat more mathematical jargon, the positive and negative signs are "fundamental operators," which indicate certain operations to be carried out. The number associated with the operator designates the extent of the operation.
The plus, **+** . positive, sign has its initial use in addition in the sense of increasing a pile, of no particular dimensions, by a certain amount described by a counting number written after it. The minus, -, negative, sign represents the opposite operation of removing a certain specified amount.

In Physics, the positive sign often represents a "charge" associated with the proton, and other species having this type of commonality with the proton. The negative sign represents and opposite charge associated with the electron and other species having a characteristic in common with the electron. In this usage, the signs do not represent reversed operations but characteristics which are considered opposites.

A third usage shows up in mathematics where the signs are associated with counting numbers to form sets of "signed numbers" which seem to be able to be added, subtracted, multiplied and divided as if they were counting numbers.
However, this turns out to have problems when one does multiplication and division processes. What is overlooked is that the addition of the sign to a number gives it both a magnitude and an implied direction. Something having magnitude and direction is not a true number, it is a "vector."
A signed number may represent movement away from a zero point. It may also represent a line segment,.the area of a surface, or the volume of a three dimensional figure or some "higher order figure" depending on where it occurs in a sequence of operations.
 Signed numbers are usually handled according to a convention wherein the positive sign is considered as representing motion to the right of an origin, upward from an origin or forward from an origin. If one multiplies three positive signed numbers together, say plus two times plus two times plus two, (+2 x + 2 x +2) what one has really described is moving two units to the right of the zero point, moving this "two-units-line" (actually a new "starting point") upward to form a square, then moving this square two units forward to create a cube which is situated to the right, above and in front of the origin point. This "eight-cubic-units entity is called a positive volume because we say "+ x + x + = + " as we consider that the + sign represents travel " in the same direction" while the negative sign represents travel in the opposite direction, reversal of direction. This signed unit, like the line and the square, has a direction associated with it which would be at right angles to the last motion. Labeling this unit as positive continues the vector content in accord with the direction change upon which it is based. As we have seen above, each operation represents a change in direction, but of 90 degrees, not a reversal. If we go, + 2, +2, -2, in our sequence of operations, we will go to the right first, up second, and back from the "center-plane" third to form another eight unit volume which will be above, to the right, but behind the point of origin. This will be considered a "negative volume" purely by convention as it has one negative sign associated, however, this convention does preserve the vector designation by the conventions observed.

(Note that we have used the negative motion in the sense of a "third negative direction motion" as it was the third operation. Considering it as "first direction negative motion" would have not created a cube but moved the square to the left of the "y" axis, not valid as a multiplication.)

One factor that seems to cause confusion in using signed numbers is the fact that the positive and negative signs are considered to be associated with change in direction form a starting point with each reversing the other. However, in the usage of the signs in multiplication, the reversals are not 180 degree true reversals but 90 degree, right angle changes according to a standard pattern.

As the positive numbers are successively associated with "positive values," right--as in handedness, upward--toward the Heavens, and forward--"progress." Negative numbers are associated with the reverse, left--"sinister" or left-handed, downward, and backward. The operation, -2 x -2 x -2 , would create a cube, which was to the left of the origin point, below the 'origin line," aka, the "x-axis" and behind the "origin plane," aka, the "xy-plane." Note that the first square formed would be considered a positive number as it is "minus x minus = plus" but the third operation, adding another direction considered "minus" labels the resulting volume as a "negative number volume."

Summing the above, a signed number represents a line, two signed numbers multiplied together represent a plane and three signed numbers multiplied together represent a volume. The order of the multiplication process will determine what plane or volume is described. The operation," +2 x -2 " represents, by the conventions used, a square which is to the right and down from the origin, while the reverse operation, "-2 x +2" forms the representation of a square which is to the left and above the origin.

It can be seen then that while 1 x I x I as counting numbers still represents the original one "pile." Plus-one times Plus-one times Plus one represents one whole. However, it is one whole cube, one length to a side, not one continuous line.

Similarly, it can be seen that the cube root of eight as a counting number is simply the number two. The cube root of +8, as a "vector cube" has the "absolute value" of 2 but this two can be either a positive or negative vector depending on which of the "generating sets" it belongs to and the order in which it falls in the set. A positive-volume-vector, +8, can be generated by any of four sets of three "signed twos." These sets are as follows: {+2, +2, +2}, {-2, -2, +2}, {-2, +2, -2}, or {+2, -2, -2}. A negative-volume-vector,-8, can be generated by any one of the sequenced-operation sets {-2, -2, -2}, {-2, +2,+2}, {+2, +2,-2} or {+2,-2,+2}. Considered this way, assigning a "signed-root" to a signed number, is a difficult and tricky business which would require a knowledge of the history of the signed number in question!   It is no wonder that the mathematicians seem to ignore "odd-number" roots of signed numbers and consider that the square root of minus one is "plus or minus **i**, an imaginary number; which, truly it is. In the most basic unit of "minus one" we are considering a line vector of a unit length. The question arises, " How does one take a root of a line vector running backwards?" Actually the root would have absolute dimension of one, either plus or minus, as one would have to be speaking of the "second-order-vector-square" which can be generated by either of the sets, {+1, -1} or {-1, +I}. Mathematicians have no trouble with saying that the square root of +1 is plus or minus one as it is generated by the two sets, {+1, +1} and {-1, -1}, sets within which the internal values appear to be identical to one another. As one can see from previous discussions that the internal elements are not identical but represent different directions of the vector depending on their position in the sequence.

The use of the two signs with different meanings of operation, reversal, or direction causes some interesting problems in understanding mathematics.

The Operators as "Exponents."

Another usage of the operator signs is in the useage as exponents, signed numbers written to the right and above a "base" number which represent in the case of a "positive exponent" the number of times the base multiplies "one" or, in the case of a negative exponent, the number of times the base is divided into one, Hence, any number with an exponent of Zero, neither negative nor positive, represents the simple number "One," neither multiplied nor divided by the base number.

In working with numbers, one often encounters fractional exponents, "roots." These are, as we remember from grade school arithmetic, are numbers which multiplied by themselves the designated times will give the base number, e.g $9^{0.5} = 3$. The rules we learned in grade school work well when one is using "absolute values," unsigned numbers; however, it runs into complications as soon as one starts to operate with signed numbers. The problem probably arises from the inherent fact that absolute number values do not have a directed motion automatically assigned to them; whereas, signed numbers do. A positive number is associated–usually–with a motion upward, to the right, or forward, with a negative number associated with motion downward, to the left, or backward.

When we take the square root of four, we realize that it is the number two, when we place two units down twice we get four. When we are working with signed numbers we have a different situation.

If we are taking the square root of +4, we are actually asking the question, "What is the directed side length of a square which we consider to have the area 'Positive Four' when we operate according to the conventions associated with signed numbers?" By those conventions we can see that both +2 times +2 and -2 times -2 fit this criterion, so we say, quite correctly, that the square root of +4 is either +2 or –2, Perhaps we would, however, have been more accurate in saying: "There are two sets that could define the square root of the signed number, +4, the set, {+2, +2} and the set, {-2,-2}".

The reason that this last was said will become clear when we discuss the situation for the "square root of -4." Let us analyze this problem as we did above. The question we are asking "What two directed numbers will produce an area which by our conventions of directed numbers will be assigned a value of -4?" This occurs again in two cases, producing two sets, {+2, -2} and {-2,+2}. As these are directed numbers the set, {+2,-2} is not identical to the set {-2, +2} as they represent opposite directions of sequential motion. With the "Positive Area" situation we found that we could create what we called a positive area by going in a positive direction then turning in another positive direction, or going in a negative direction and then turning in a negative direction again. For a negative area we can start out in a positive direction, then "turn negative" or start in a negative direction and "turn positive." By this analysis, the square root of "Negative One" is not an "imaginary number," per the usual mathematical solution for the situation; but rather is again a situation of two directed sets, however, these are not of the same directed number, "operating on itself," which allowed us to say, in the other case, "Plus or Minus One" is the root, implying each directed number "operating" on itself, but "Plus and minus one" the two operating on each other.can be clearly seen to be two differently directed,units.

The concept of imaginary numbers arises because of the ignoring of this fact of the directed action factor inherent to signed numbers. The concept of directed sets as roots can be extended to higher roots. For the cube root of +8, one may write the sets, {+2, +2, +2}, {-2,-2,+2}, {+2, -2, -2}, and {-2,+2,-2}. Noting four sets that can be considered the "cube root" of +8. A similar group of 4 sets represents the "cube root" of -8. A fourth root would presumably continue the pattern developing, eight sets of 4 units each. This is left to be proven, or disproven, by the reader. The idea of an endless set of imaginary numbers as successive even roots of "Minus One," is an interesting concept; but, by the above analysis, appears to be based on a misunderstanding of the significance of signed numbers. The use of a signed number indicates a motion in a direction and can be considered to define a "dimension."   The original two articles combined and edited above were Posted on SciScoop   March17, 2009.

Another sign that we sometimes take for granted is the equals sign, **=.** This usually is considered to mean "equal in amount" and handled as such. Again there is a subtle distinction here whether we are working with absolute values without designation as to direction or signed numbers which have an implication as to direction with respect to a set of axes, usually the Cartesian Axes. That is, the familiar **"x, y, z,"** set of axes. The use of absolute values does not imply a balance about a point in a specific direction. The use of signed numbers does. Use of absolute values implies a balancing in space but does not imply a specific point nor a specific direction.

Sequence in Science and Math. The Various "Time" Concepts

A very important concept in science is the concept of motion and the consequences of motion, sequence. The two are intertwined in thinking in a concept called "Time," which, unfortunately, has been extended and misconstrued to the point of causing much misunderstanding. An example of this would be the comment made by one correspondent who said, "You need at least fifty years experience and PhD's in math. And physics before you would dare to write on such a deep concept as time!" Apparently this was his set of qualifications. In my opinion, the idea that "Time is a 'deep subject," simply emphasizes the fact that the word has come to mean a number of different things. The original concept of "Time," can be said to be the measurement of sequence in relation to some reproducible natural sequence so that sequences can be coordinated and controlled. Other meanings hinge upon the idea of time as a instant in any sequence. The speed of light is the natural ratio of distance measured by comparison to some physical object to distance measured with respect to some natural sequence. The following article, written about 2008, discusses some of the ideas of "Time." In fact, it was probably the article that was being presumptuous published in the criticism of the person quoted above. It was first published on Helium.com. It is reprinted here with some additional comments.

Essays: The concept of time

There are a number of Essays on the concept of "Time" to be found under this title on Helium.com. The following short essay was rated "22 of 42" at the time it was copied into Google Drive a few years ago. This article reflects a simpler, more restrictive, view of "Time" as compared to, for example, "Minkowski Time" which seems to extend the idea of "Time" to a "Dimension which encompasses all possible sequences stemming from any given starting point. " *(This is the writer's own interpretation of Minkowski Time, and is not necessarily Minkowski's intention.)* Time is a human construct used to keep track of sequential motions in space, it is sometimes considered as if it were a fourth spatial dimension. Time is usually considered to have three aspects, the past, present and future. These can be considered to correspond to motion of a specific point on a wave front. The past is the total sum of all motions that led to that instantaneous position which we consider the present, and the future is where that wavefront spot may be considered to go in the next instant and all the instants which may follow present is the result of a certain sequence of motions, which we call the past, something which no longer exists, but which is nevertheless a "fixed construction." To go back into the past, in a physical sense, would require a retrograde repetition of all of these motions, a set of motions which would increase instant by instant in a huge geometric progression. Our "wavefront" would have to move backwards in a perfect reversal of the sequence by which it had previously moved forward. Even were this possible, one can see, that, since the direction has been reversed, what was "back" is now "forward." The wavefront, while "retracing the past," would actually by "moving into the future." In trying to go into the past, a "time-traveler" would actually be attempting to create a future which was a reversal of the past. If one considers that it may be possible that long wave fluctuations in the "Matrix" in which we exist creates multiple adjacent universes in which certain sequences may coincide, it might be possible to move from one alternate universe into another which would correspond exactly to some point in one's own past.

There is no real indication that there are such alternate universes; and, were they to exist, it is highly unlikely that they would correspond in such way as for there to be possible entry from one to another.
The above was written about 2008. It does not note other usages of the idea of Time such as "Psychological Time," the apparent "speeding of time with time with age," that is the experiencing of a sequence as a proportionate to the total past sequence.
 It also does not go into the fact that time, measured in reference to some reproducible, natural sequence, can be used to measure space, and vice versa. It is also not pointed out that the natural ratio of sequence measured as a natural cycle and space measured with respect to some physical object, is "c", the Speed of Light.
 This fact appears to the basis of the apparent validity of Minkowski-Einstein Space Time, which could be considered a "Mathematical Substance in continuous Motion". Space would represent the universality of the dot matrix which automatically fills mathematical space whenever any mathematical operation is carried out, and "Time" always implies sequence. That is the presence of the concept of Time indicates the possibility--in fact, the certainty-- of motion, Hence Space-Time is a mathematical representation very nearly equivalent to the "Substance," of the "Oscillators in a Substance Model."
It may be noted that the Minkowski-Einstein Space time Model which postulates four "dimensions of existence," three of one sort and one of another could probably be also expressed in terms of three dimensions of "Time" in one inclusive dimension of space.

Equations and Differential Equations

Continuing comments about some of the concepts which may cause difficulties in considering mathematical models as "Reality," rather

than useful, possible approximations of some part of reality, let us consider the equations that are fundamental. One of these is the equation, $v=at$. Velocity equals acceleration times time.
 (Note again the multiple usage of the word, "time.")
This is used without thought of the fact that in any real situation in which a velocity was being measured, some physical object, having "mass," would have to be moving with respect to a point of reference in order to measure the velocity. Therefore, there is an implied basic unit of "mass." involved somewhere in the situation. This may seem a trivial observation, but notes the difference between a mathematical operation and reality involving observation.
Another equation is the definition of momentum, "$p=mv$." Momentum is considered as the rate of change of "Energy" with respect to "Time." (As "Time" is the standard term for "set of sequences" any change may be considered as being with respect to time.)
Then there is "Energy," the capacity to cause change, usually meaning Kinetic Energy, the motion or change that can be expected to happen when a moving object collides with a fixed object, $E=mv^2/2$. We have defined a number of ideas. Or have we? We have mathematical definitions, and we have managed to involve the term, "Mass' in all of them without actually finding any real definition for mass other than perhaps, "an aspect of something which has existence."
If we switch from mathematical definitions to a concept from the Oscillators in a Substance Model, the idea of some sort of all encompassing basic substance of unknown extent and undefined basic unit," (O.K., O.K. a basic something that is made up of some smaller basic somethings), we can develop some useful concepts immediately. The concepts of pressure, force and mass. Pressure is quite intuitively understood as a push. In an all encompassing something there would be an average, almost constant push. A one "Force which was its own equal and opposite." This is somewhat misquoting the standard definition of a true "Force" as having an "Equal and Opposite." From this we can find a definition of mass, as a measure of the Force within the surface of some unit within the substance such as to allow that unit to have existence.
Mathematically, Force is mass times acceleration and Pressure is Force per unit area. $F=ma$ and $P=F/A$.

What mathematicians know, but do not usually comment on, is the fact that any number of aspects of a situation can be ignored if they are constant, They can be divided out, replaced by the number, "One." Conversely, some aspects of a situation may always be there and be ignored because they are constant. It is probably quite useful to consider that within our universe there is some sort of a basic acceleration, so that for many purposes, we can use mass and force interchangeably. Similarly, in working with velocities, it, also, may be well to remember that, in real circumstances of measurement there has to be at least one basic unit of mass involved. Also, as Einstein would probably be quick to point out, any operation involves the transfer of information hence the fastest known velocity of information transfer the Speed of Light, may need to be taken into consideration. "Queen Mathematics" does not usually not such things. Physicists ofthe consider that any legitimate theoretical work must be defined in terms of "differential equation," that is equations which involve the consideration of "infinitesimal, positive changes." Note the operational word, "positive."

One blatant case of nonsense involves a constant called, Dirac's constant, or "h-Bar," written with a bar through "h", the symbol for Planck's Constant. For the moment, let us say simply that Planck's Constant has a per cycle value, Dirac's Constant divides that constant by 2 Pi, which, somehow, to scientists seems to, somehow, "normalize" the constant. Actually, what is happening is, if the cycle in Planck's Constant were to be measured in "Degrees of arc," Dirac's Constant would be measured in "Radians." For most mathematical purposes, nobody really needs to worry which is used. Any way Dirac's Constant can be considered to have to dimensions of Energy multiplied by time, somewhat ignoring the 2 Pi, and one sees the equation written, $E \times t = h\text{-Bar}$, In mathematical terms, Energy times time is considered equal to a constant. To the writer, this equation is not valid, questioning the validity of "h-Bar." However, the situation gets worse.

What then is written is Delta E x Delta t = h-Bar. A Delta is a small, positive increase. an "increment.", This appears to be a legitimate equation. However, a small positive increment of each of the originals will not have the same value as the original variables, this equation, therefore, is flawed, and the shift to infinitesimals by, writing $dE \times dt = h\text{-Bar}$ is therefore nonsense. $dE \times dt = h\text{Bar}$ is a "differential equation" and any use made of it will lead to nonsense conclusions.

The convention of considering any number without a designation as being a positive signed number may be one of the most accidental dangerous tricks in mathematics as applied to real situations. As discussed earlier, the operation in signed numbers leads to all operations being considered automatically in the "right-upper-forward" octant of a set of x,y,z ("Cartesian") coordinates. This would appear to have the possibility of ignoring ⅞ of a model. The danger inherent to considering only "positive incremental" and "positive infinitesimal " change was seen in the "false differential equation" situation discussed above.

The Oscillators in a Substance Model's "Silly Little Equation"

The Oscillators-in-a-Substance Model considers the units which make up space to be rotors having an average tangential velocity of rotation of the Speed of Light, "c," Looking for further characteristics of these otors, or of other units of space, the Model takes the position that the ratio of Planck's Constant--a constant of nature related to light energy--and the Speed of Light would be another very fundamental constant of nature. This constant,   "h/c," has the dimensions of mass times distance. Applied to a rotating unit this could be mass time radius. This is "angular momentum." The dimensions, mass times distance, are the dimensions of "Work" and "Energy." The observation that this would appear to be a constant which is not time dependent, that is , it is an independent quantity, a "quantum." Therefore, its value, about 2.21×10^{-37} g.cm. , would appear to be the "Quantum" which was never despite giving the name to the "Quantum Revolution of the XX Century."
The equation which arises, $m \times r = h/c$, is an equation of the form, $xy=K$, a fundamental equation of mathematics, which has a longer form , $xy=K=yx$. In This form, the equation type could be considered to represent symmetric oscillation. Here is where

this model (OS) trusts the mathematical ideas to mirror reality by postulating that if one limiting value of mass or distance can be determined, then, possibly the corresponding limit can be determined, and another, complementary set of limits be determined by switching the absolute values.

Working with absolute values, the 'rest mass" of the electron was assumed to be a limiting mass and divided into the value of h/c, the resulting value for a possible radius was found to be known in the literature as the Compton Wavelength. Using the rest mass for the proton produced the same type of results.

This leads to the interesting speculation, assumption, that the equation, m x r= h/c could define a basic oscillator family of existence.

(When essentially the above was told to a group of people from a certain internet site who claim to have physics background, the result was derision, "Your silly little equation…." Asked why they said it was a "silly little equation," the answer was, "Because it is a silly little equation! "

(Remembering that Mathematics, the Queen of Sciences," can be a misleading trickster, I have to concede that the assumptions made in the use of the "silly little equation"may be wrong. However, accepting the mathematical premises does provide solutions for a number of "unsolved mysteries of physical science. ")

An additional assumption made about the equation is that the square root of the constant, h/c, i.e. $h/c^{0.5}$, may be considered as an average value of both the mass and radius of the unit under consideration and will be the average for all of the members of the family, possibly even the defining values for a basic "control oscillator." This value is 4.7 x 10^-19 grams or centimeters.

An essay that has been previously published in the book, Off the Wall, Vol.1: Ideas to Consider" is reprinted here, as another view of what happens when we aren't careful about how we handle our thinking.

Modern Physics Mythology: Fantastic Theories of the 20th Century

Summary. *This essay presents arguments that a number of ideas which are commonly accepted in scientific circles are in error for reasons of reaching conclusions from inadequate consideration of data, erroneous preconceptions. or both.*

The Twentieth Century was a period of great technological and scientific progress. However, it seems, on recent examination, that some of the advances almost took place in spite of the theory rather than because of it.

The theory problems might be said to have started with the Michelson Morley Experiment of 1890 which was a technological masterpiece designed to test a particular theory of an Aether in which Matter moved, or which was moving about Matter. The result of the experiment was that the Speed of Light was the same--at least within experimental error-- no matter what the direction of motion of the transmitter and receiver with respect to one another.

This was interpreted to mean that the "Aether" did not exist, the space through which matter moved was empty; and, somehow, light was carried through that empty space by "Photons." A definite case of "jumping to a conclusion which was not truly warranted." Instead of considering what this result might have actually implied about what might be the components of space, the conclusion was that "Space was a void," because the results were not consistent with the preconceived notion.

This tendency to reach a conclusion on insufficient data. or as the result of an erroneous preconception, received an additional twist when Albert Einstein came upon the scene some fifteen years later. Einstein's work could be said to have started a new kind of science attitude which prevaded the rest of the Century. Einstein is said to have remarked that it made no difference whether space was empty or full, assuming it was empty was mathematically easier. Apparently he was mathematically correct. When one does anything with mathematics., they automatically fill their mathematical space with a dot matrix.

[It seems to produce a more workable view, however, if Reality be considered to be within a filled space, a space filled by something resembling a substance at its triple point.]

Einstein, also, is supposed to have said, "Mathematics is the Reality."
This attitude seems to have led to a great deal of the theoretical work which came out of the next Century to have a strong "reverse bias" with respect to mathematical theory development and data.
 Before Einstein, the procedure seems to have been "study a phenomenon, collect data, develop a mathematical model, see what it predicts and check to see if the prediction holds true."   That is, the theorizing was "data driven."
 More recently, many scientists seem to feel that they can shortcut the process by developing the theoretical model first, then, looking carefully for data to support that model. .If one bit of data can be interpreted as to support the model, the model is assumed to be proven, and any other data is carefully interpreted to fit. A prime example is the "Black Holes in the Centers of Galaxies Concept ";believed by astronomers , which is so carefully espostulated by an astronomer, Dr Alex Filippenko, in a course from the Great Lectures Series, "Black Holes Explained."

Black Holes had their apparent inception in the Ideas of a man named Schwarzschild who interpreted some of Einstein's theorizing to claim the existence of such a phenomenon. Among the thoughts was, apparently one that these Black Holes would appear as very massive objects. It was found that suns near the Centers of Mass of galaxies were moving in such a way as to apparently be orbiting a massive object. That object wa taken to be the anticipated Black Hole.

What no one seemed to realize was that all the mass of a diffuse, but coherent, unit--such as a Galaxy-- would sum its mass to a point at the center of balance just as does any other unit.

Yes, there is a point of tremendous, apparent mass at the center a Galaxy. It is equal to a point summation of all of the separate masses involved in the galaxy; but. it is not a "Black Hole."
The phenomena observed are attributable to what happens in the region around the center of this summation of vectors. There is neither a physical point at that position nor any infinitely deep hole.

Black Holes are a part of the modern mythos generated by the idea that it is simpler, easier and just as valid, to do "Theory Driven Dataing" --which, of course, is never called that--rather than the "Old

Fashioned, Data Driven Theorizing," a term which is never used, either.

.

the Standard Model of Particle Physics--so eloquently expounded in Dr. Sean Carroll's Great Lectures Series Set entitled "**The Higgs Boson and Beyond**--" is a prime example of this type of thing. The "Standard Model" is possibly the most honored, most expensive construction of its type known.

In my opinion, the Standard Model of Particle Physics is the most misguided, pseudo-scientific "House of Cards" ever constructed.

The original inception is in the idea that the basic units of atoms could be released by smashing atoms into each other at very high velocities.This has been likened, perhaps quite accurately, to trying to learn how cars are constructed by dropping them off cliffs.

What the scientists did not seem to realize was that. instead of releasing whatever was within atoms, they probably would, instead, create new and different things in the fragments that they released.

About the time that the number of supposed basic units became unwieldy, there was a paper which proposed a unit called the Quark.

This writer has always been of the opinion that the original paper proposing Quarks, published straight-faced as if it were a legitimate theoretical idea, was almost surely a clever satire on the tendency of scientists to make their theorizing a complicated as possible.

 Apparently that opinion was not shared by the physics community, which seems to have happily adopted the ideas, found some electron scattering data which seemed to note a 2:1 scattering ratio for electrons,and promoted Quarks from there being three, as originally proposed, to being six.
The particle physicists after defining various, "basic" units decided that they still needed one even more basic background unit which became known as the "HIggs Boson."
 This fantastic theoretical structure known as The Standard Model became so respected--bolstered by a number of Nobel Prizes-- that there was constructed the biggest engineering science project ever, the Hadron Collider constricted by CERN under the border of France and Switzerland. This project, costing, untimely, billions of dollars, had a major purpose of finding the Higgs Boson.

Finally, in 2012, it was announced that the HIggs had been found. All was now right with the science world, It was now known how and why everything worked.... Higgs and a still living coworker--one had died--received a Nobel Prize for their theoretical work.
 The writer of this is quite sure that the "Higgs Boson" was, actually, if quite accidentally, the "Biggest Hoax On" (themselves), ever carried out by well-meaning scientists.

   Dr. Carroll, in his lecture series, points out that every collision in the Hadron Collider generated more data than could ever be housed in all of the databanks of Earth so "...this is "where the theoreticians came in (to tell what to keep and what to throw away, where to look,)"
 In any other scientific field, this would be definitely frowned upon as drastically "Cooking the data."

At long last something was found whcn could claim the right place. They had searched at least two other region which were expected to be more probable and, apparently, this was the last chance, Something had to be found to justify the billions of dollars expended, so something was found, and said to be what was being looked for.

One may cynically suspect that wishful thinking may have clouded judgment.

The HIggs Boson and the HIggs Field, could be considered to some what parallel the idea of a "Basic Substance" and a "Central Control Oscillator," which arise in the much simpler "Oscillators in a Substance Model" However, the idea that this little unit, which can only be observed under the most drastic of conditions "bestows Gravity," is conceptual nonsense.

In the Oscillators in a Substance Model, there is a very simple, logical definition of Gravity as the commonest manifestation of the "One Basic Force of Nature," that force being a very slightly varying pressure throughout the "Substance of Existence." **Gravity is a simple result of the fact that between any two entities there will always be less of the Substance of Existence on the line between them than on the line on the far sides of them**.

The search for the Higgs Boson almost certainly will, someday, go down in History as the most expensive scientific experiment ever carried out on the basis of flawed logic. A much simpler model of existence --the present form of which is now called Oscillators in a Substance Model. This theory could have been very easily developed about 1910, or earlier, as the data was available in the work of Mickelson and Morely and Max Planck.

The venerable, and useful, proton-neutron nucleus model of the atom, also, apparently has its base on very little actual information.

 It was known that the mass concentrated in the nucleus was greater than that attributable to protons. Therefore it was logical to assume that some part of the atomic nucleus was neutral in charge. This was attributed, without question, to some set of neutral units. **When the neutron was discovered, By Chadwick in 1932, it was found that this might fit the neutral unit which was needed. It appears that no one ever looked seriously at any other explanations**.
 The writer gives other possible explanations in two books published in 2015. None of those ideas are necessarily perfect, but indicate alternatives to the current complicated and confusing, Proton-Neutron-Strong -Nuclear-Force-Weak -Nuclear-Force Model which is currently in vogue.
Whether, from getting a model first then taking one piece of data as confirmation, or taking one piece of information and considering it as proof of some

"obvious conclusion," physics theorists have created a shambles of tangled theory. Another piece of the case is the Matter-Anti-matter Annihilation Belief which has its base in the fact that if an electron be placed in contact with an anti-electron, the two will vanish with the release of energy. The same apparently happens with the proton-antiproton set. Therefore, the scientist say, "Antimatter and Matter annihilate on contact QED." There have been many very learned--and often very abstruse--papers written based on these ideas,   much discussion as to why we do not observe anti-matter.

 In a little article printed in EGO Out, the blog of Dr.Peter Gluck, in Oct. 2015, this writer pointed out that mirror images combine rather than annihilate. It also can be pointed out that interaction of Hydrogen and Antihydrogen is a lot more complicated than simply the interaction of the components. Also, it is known that **e-** and **e+** combine to form a "compound" dubbed "Positronium" which has lifetime enough to have a chemistry, before it disappears. It makes sense that Positronium is an analog of molecular Hydrogen and the combination unit, called "Zerotron" by this writer, is the analog of the Hydrogen atom.

The idea that there is an unnoticed, rather ubiquitous unit which results from the union of the electron and antielectron, and can be split back to them explains both "annihilation" and "pair production." Also, the existence of such a unit, making up some 95% of the particles of existence, would explain rather neatly, perhaps too neatly, the "Mystery of Black Matter." Without going into the arguments justifying the opinions, the writer suggests that the electron and the antielectron are interconvertible and, indeed, each spends 40% or so of its apparent rotation-inversion-inversion-of-rotation cycle as the opposite unit. It, also, may be argued that one may model an atom as made up of a set of electrons and protons, and another, unnoticed set, of anti-electrons, anti-protons. The latter set being considered as "neutrons," in conventional thinking.

Still another model, not inconsistent with this, but different, notes that almost all known atoms may be considered as combination of Deuterium, Tritium and Helium 3, or, alternatively, nuclei may be considered as formed from a three-unit -(of mass)-with-double-positive-charge, a three unit with a single positive charge; and a two unit with a single positive charge.

Now, if in the first unit noted above there is considered to be one "embedded" electron, in the second unit above, there be two "imbedded electrons" and in the third unit above, there again be one "imbedded electron" the number of "embedded electrons" turns out to be the same as the number of neutrons, or the amount of anti-matter, depending on which view one wishes to take.

Hence, there are several possible mathematical models of atoms which may be utilized. It is not claimed that any are "exact truth," but each has possible utility for understanding one process or another.

For instance, "Beta negative" emission may be understood as "Tritium unit conversion to He3. unit within the atom." Conversely, Beta plus emission, or the equivalent K-electron capture process, may be considered the reverse, conversion of He3 to T.
Alpha particle emission can often be rationalized as combination of two Deuteron units at an Alpha within a dication. This suggests that Alpha

emission may be a function of the dication of a particular isotope. This fits well with the known collapse of Be8 to, eventually, two Helium Four atoms. This is a process that one would not expect to take place with neutral Be8; but seems very logical for Be8++.

It may be argued that the only reason the Big Bang gets any credibility is that physicists never bothered to truly define Mass or Energy nor define what "Matter" be. Otherwise, it would be realized the idiocy of considering all maffer to be reducible to one tiny dot which would explode in one big "Cannon Shot, a Big Bang". A model which seems closer to explaining the current situation seems to be to start the process with a "Tiny Squeak," which still continues, a sort of cosmic Machine Gun which is still firing, rather than a Cannon which fired once.

The Origin Myth which the Oscillators in a Substance Model of Existence creates, postulates a Control Oscillator operating at about 6×10^{28} cps. Coming, somehow, into existence within a paleo-substance causing organization of that substance into proto-units, probably the unit dubbed the Zerotron by this writer. These units are shock-wave-distortable to neutrons which collapse to electrons and protons and /or positrons and antiprotons. This origin myth seems consistent with an expanding universe, the known microwave background pattern, and other factors of the model from which it originates.

By this Origin Myth, the vast majority of matter would still be clustered near the origin and would thin out throughout the lobes, in one of which we exist.

The idea of a paleo-substance is essentially the same as the idea of a Substance at its Triple Point of unknown Extent and Basic Unit which is fundamental to the "OiaS" Model. Which leads immediately to the idea of a "One Force," a ubiquitous, slightly varying pressure, and to an easy definition of Gravity as the "Apparent Attraction" between any two objects having mass because there is always less of the Substance of Existence between them than there is or the outside. This concept of a Universal Force also leads to a definition of Mass as the force within a unit, expressed against the rest of Existence, which allows that unit to exist. In other words, <u>"Mass" is a necessity for existence.</u>

[An aside comment: Dr.Carroll in his lecture series says something to the effect that existence is because of the

presence of fields and that particles are what are found when fields are examined. My problem with this is that I see Dr.Carroll himself as a collection of particles and find it inconsistent with what he is saying that he exists to be able to believe his theories. In other words, what are the examinations that bring him (or me) into existence?]

The guess as to the frequency of the "Control Oscillator comes from the idea that in order to carry the wave disturbance of light, the units of the "void" need to be rotors with a tangential velocity averaging the Speed of Light. Since the work of Max Planck also deals with light and its relationship to Energy, it was realized that the ratio of Planck's Constant and the Speed of LIght would be also a constant which has the dimensions of mass times distance, that is "Torque" or "Work." This statement in mathematical form is m x r (radius of a circle or sphere = h(symbol for Planck's constant) /c (symbol for Speed of LIght.) m x r = h/c was later noted to be a "time independent work function, which would be the amount of work involved in one fundamental cycle of Electromagnetic radiation" therefore, it appears to be the elusive "Quantum" which gave the name to the Quantum Revolution of the Twentieth Century.

From basic algebra, $m \times r = h/c = r \times m$. That is, if one set of limits can be found, a guess at another set of limits can be found by changing the absolute values around. If one does this with data from the electron and the proton, considering the rest masses that are found in the literature as being limiting values, something very interesting happens. Both the electron and proton fit into this equation, with the "radius" identical to the "Compton Wavelength." If one turns the absolute values around one finds that apparently the electron would be both larger and smaller than the proton and heavier and lighter, depending on which limit one was considering. In addition, it can be suggested that both have the same average value, which would be about 4.7×10^{-19} Grams at 4.7×10^{-19} cm. , that is $(h/c)^{0.5}$. The estimate of the possible frequency of the postulated Control oscillator comes from this value, taking the "leap of faith" to consider all basic units as "Encapsulated Wavelengths" of the Control Oscillator.

The reason that the result of this is being called a "Creation Myth Created by the Oscillators in a Substance Model" is that it evolves from speculation from a mathematical model. Even though the mathematics is very basic. This is perilously close to the jumping to the conclusions that are being criticised so strongly in the rest of this essay.

Until there is a good deal more evidence, the Oscillators in a Substance Model, needs to be accepted as just that, a model which seems to simplify and unify a great deal of information. However, it is only a model, and will be expected to undergo modification or even refutation as information develops.

Most of the above essay is based on the Oscillators in a Substance Model, hence the material is covered in more detail in the books that are devoted to that model.

More About Unconventional Mathematics in Physical Science. Although the conventional thing to believe is that any physical science theorizing should be expressed in differential equations. There are science ideas that may have some mathematical expression but do not seem to fit weil into this type of expression. Particularly difficult is the problem of fitting concepts of the atom into the differential equation straightjacket. One such idea is the long-standing set and subset models of electronic structure of isotopes. When this is extended to the idea that the set-subset model,-- e.g. Helium as having a "1s2, 2s2" configuration-- represents some sort of an electron-proton inter related set matched to an unnoticed anit-electron -antiproton set, there seems no way to express the situation in "Dif-Eq." form.

The "OS" Model has additional suggestions that arise for consideration as atomic models which may add insights, but would require other mathematical patterns than Dif. Eq. One is a model which suggests that an atom should be considered as having a nucleus. This nucleus is designed by the scope of the small ranging vortex oscillators known as protons. The atom would have other defined surfaces. One far within the "nucleus," and the other far outside the nucleus, both of these defined by the much wider range of the oscillation of the electrons. (According to this model, the electrons have a range from far outside to far inside the nucleus.)

Considering the electron as a dual oscillator, the lower frequency oscillators would define the surface of the atom, with the much higher frequency other, coupled oscillators defining an inner surface far within the nucleus of the atom. This idealized, dual oscillator vortex model is handy for modelling.

Although these vortex oscillators could be variable frequency oscillators, the dual frequency idea seems easier to visualize. Considering our vortex model and the fact that is some published evidence that the electron turns 720 degrees-- that is, around twice in some manner in each cycle-- one can develop an interesting approximation of this motion by considering the central, higher frequency oscillator as being fixed at the center of a set of Cartesian ("x,y,z") coordinates and the outer section spinning about it. If we set the initial situation at spinning in the three positive directions, (right, up and forward) and call this situation , "+++," we can pretend that the outside pressure will be such that each of these directions will change progressively from + to - so that the position of the vortex about the origin may change successively through the sequence, +++, +-+, +--, ---, --+, -++ and, finally, back to the original +++. This represents two complete reversals, +++ to ---, and --- ,back to +++. The space occupied in this process would be spherical and the positions occupied by groups of these oscillators could, presumably be considered as some sort of coordinated spacing of spheres.

Following up on the above ideas one may suggest that the "shell-subshell-orbitals" model of the atom may actually be a definition of spherical packing in groupings corresponding to successively, up to two things "anchored" to a point, i.e. "s orbital configuration<, up to six units about a point, "p-orbitals," up to ten units, "d-orbitals," and up to fourteen coordinated,"f-orbitals." A suggestion might be made that the principal sets are separated by an amount corresponding to the "basic "Quantum," 2.21×10^{-37} g.cm. This, is simply a guess, in so far as the writer knows, neither this nor the above calculation has any "experimental validation." The above is all simply speculative modeling based on assumptions inherent to the Oscillator in a Substance Model (OS).

A list of some of the basic assumptions of OS is the following:
The Speed of Light and Planck's constant are valid natural constants, probably as statistical averages.
The ratio of these two constants is a constant, h/c.
This ratio has the dimensions of mass times distance.

Mass times distance applied to something spinning will be mass times radius. Mass is the natural force associated with any existing particle or wave.

An equation of the form, $xy = K = yx$ can define a family of oscillators.

The square root of "K' can define a central oscillator.

The square root of the constant, h/c, Planck's Constant divided by the speed of light is the above kind of definition. This square root can be considered as defining central values for both mass and radius.

If the absolute values of x and y define the limits of an oscillator, then switching the coefficients probably will define a congruent limit.

Mass times distance is a "work function," which is independent of the time involved to do the work, hence $h/c, 2.2l \times 10^{-37}$ g-cm would represent a basic work/energy function of nature. This quantum may define the differences between what are called "electronic energy levels."

It seems fairly clear that the OS Model rests on some basic assumptions as to how mathematical reasoning may apply to reality: however, there seems to be no reference to differential equation modelling.

It may be noted that Quantum Mechanics is based on use of the same two constants in a different manner which involves differential equation modelling. The OS Model has two interacting and interdependent parts, the Substance and the Oscillators which operating within that substance. The best known of the accepted models, Spacetime and Quantum Mechanics, appear to be "partial half-sections" of this of model.

Space-Time has the effect of defining a substance without reference to oscillation. Quantum Mechanics defines "wave functions" which may be considered as related to oscillators, but appears to make no mention of oscillation, nor of an underlying substance or matrix.

Something for Mathematicians to Comment on.

Very often a differential equation and its integrated summation are related as equations which we memorize. One example is the equation for momentum, $p = mv$ which in the differential form is "$dp/dt=K=d(mv)/dt$". The result of integration of this is with mass held constant is the well known Kinetic Energy formula, $KE-mv^2$. However if one integrates $d(mv)/dt=K$ at constant velocity, allowing mass to change one obtains another "Energy" equation, which is not usually named in physics, $E=vm^/$. That is, Energy at constant velocity with continued attempted acceleration equals velocity times mass squared.

To complicate the situation even more, if we simply integrate the situation from "dp/dt" we get the result of the integration as $P^2/2$ and since p equals "mv" the "True Energy Equation" would appear to be $E=(m^2v^2)/2$.

It is quite interesting that, if we now "differentiate," this last expression back to find a change of mass at a constant velocity of 'c,' we get an interesting definition of Einstein's famous equation, mc^2, as the change of mass with time, if velocity is held constant at a value of 'c.' It appears that this equation may not apply to the change of all mass to "Energy" as it is generally assumed to mean.

One can continue integrations to define other things, Work or Energy expended over time equals, perhaps, accomplishment, or Energy Dissipated, this would have the form, if we consider all energy as Kinetic Energy, $(mv^2)/2$, with mass always constant, the formula would be $(mv^3)/6$, and if one went to some other dimension of accomplishments somehow compiling to different accomplishments we could obtain $(mv^4)/24$. This sort of integration could be continued indefinitely. These are legitimate mathematical actions, what they might mean in physical science theory is left to the reader to decide. The writer, at this point is clueless.

More on Models of Atomic Structure

Since the middle of the 1930s, the nucleus of atoms has been considered to be made up of protons and neutrons. Now the OS Model suggests intriguing other models. One is that the neutron count actually is a count of the average antimatter content of an atom. That is taking the view that atoms are made up of both matter-- electrons and protons--and antimatter, positrons and antiprotons.

 Even more intriguing is the possibility that each unit actually switches between Matter and Antimatter forms. Using this view, it is possible to consider that the electron orbital models which are used to correlate "outer electron" can be used to correlate "inner electrons ," or positrons. An excellent example is Natural Gold, atomic number 79, atomic weight 197. The neutron count of 118, if worked out by the standard electronic structure calculations turns out to be the nest inert gas structure. Giving a neat explanation for the stability of this isotope.

Another model that arises from OS is a nuclear model which could be called the Four Basic Units Model which fits for all but a few very rare isotopes.   This is a model of the nucleus as made up of combinations of units of mass three, charge two, "3++," mass three charge one, "3+," mass two, charge one, "2+," and mass one charge one, "1+". Using this notation of considering nuclear units as being cations, it may be noted that the "Two and Three" units may be considered to have "imbedded" negative charges, what might be called "resident electrons." It turns out that the number of these "resident electrons", or internal electrons, is also the "neutron count." Using natural Gold as the example, and listing the supposed nuclear content in the order of 3++, 3+, 2+,1+, Au79/197 can be coded as 13, 52, 1, 0. The fact that 52 is a multiple of thirteen and that there is a switching from matter to antimatter, the one 2+ unit could be considered a "balancer." This, also, may be argued to additionally help account for the stability of this particular isotope.

An interesting coincidence is that Gold is often considered as associated with the Sun. The coding sounds surprisingly like a rather decent solar calendar of 13 months, 52 weeks and an extra holiday left over.

Closing Comment

It is hoped that this little essay will furnish food for thought in looking at simple mathematics and the interaction of mathematics with reality. Additional.information that might be of interest might be found in the books, by this writer which ar4e listed on Amazon /-CreateSpace..

Thank you for reading this.

Hasta la vista.

www.ingramcontent.com/pod-product-compliance
Lightning Source LLC
Chambersburg PA
CBHW080613190526
45169CB00007B/2997